W9-CGM-968

At Home with the Beaver

THE STORY OF A KEYSTONE SPECIES

Dorothy Hinshaw Patent
Photographs by Michael Runtz

Splash! The soft buzzing of bees and chirping of birds is interrupted by a loud *WHAP!* Ripples radiate across the water from the spot where a beaver surfaced, slapping its tail before diving down into the tea-colored pond.

If it weren't for beavers, this pond wouldn't exist. An intricate wall of logs, sticks, and mud—a beaver dam—holds back enough of a creek's water so it becomes a pond.

Another mound of sticks and mud rises from the middle of the pond. This is the beavers' lodge, where the beaver family lives.

A beaver mother and father built both, first the dam to hold the pond and then the lodge that became their home. By doing this, the beavers have also created a home for thousands and thousands of living things of all sizes, shapes, and colors.

Before the beavers built their dam, the creek often dried out when rainfall was slight, so water-loving plants and animals couldn't survive here. But once the pond filled behind the beavers' dam, life began to thrive. The pond's life ranges from clouds of microscopic plants and animals to insects called water striders—which skate around on the pond's surface—to frogs, ducks, muskrats, mink, and more. Because it creates this lovely wet world that is home to so many other living things, the beaver is called a *keystone species*.

MALLARD

WATER STRIDER
MINK
LEOPARD FROG
MUSKRAT

The beaver's role as a keystone species extends beyond the pond itself. The lush greenery surrounding the pond exists because water has seeped into the ground from the area flooded by the dam, providing year-round moisture for many kinds of plants. And by cutting down trees, the beavers have opened up the land to more sunlight, helping these plants grow.

MONARCH BUTTERFLY
ON JOE-PYE WEED

Many kinds of plants ring the pond with brilliant green leaves and colorful flowers. There is joe-pye weed, a food source for the monarch and other butterflies, and orange jewelweed—or "touch-me-not"—with its spring-loaded seedpods that pop when touched. Surrounding the pond are cattails—a common nesting site for red-winged blackbirds. Pussy willow—one of the beaver's favorite foods—also grows close by.

ORANGE JEWELWEED
PUSSY WILLOW
RED-WINGED BLACKBIRD
CATTAIL

BEAVER EATING WATER-LILY LEAVES

Plants not only surround the pond; some live in the pond itself. Smartweed pushes its leaves and pink flowers above the water's surface, and glorious yellow water-crowfoot covers parts of the pond like a blanket. Magnificent water-lilies float atop the water. Sometimes young beavers roll the water-lily leaves into a wrap to eat as an afternoon snack. Water-shield, another common water plant, provides a nice resting place for a green frog. All of these plants need calm water to thrive, which means they wouldn't be here without the beavers.

YELLOW WATER-CROWFOOT
WATER-LILY
SMARTWEED
GREEN FROG ON WATER-SHIELD

Life is interconnected in the beaver pond. Under the water's surface, microscopic living things, such as single-celled algae, provide food for dot-sized water fleas. Then insects called backswimmers, which use their long hind legs like oars, feed on the water fleas. A bass rises to the surface and gobbles up a backswimmer. On a nearby branch, a kingfisher peers down, scanning the pond, patiently waiting for the bass to swim by.

Food Chain

KINGFISHER

SMALLMOUTH BASS

BACKSWIMMER

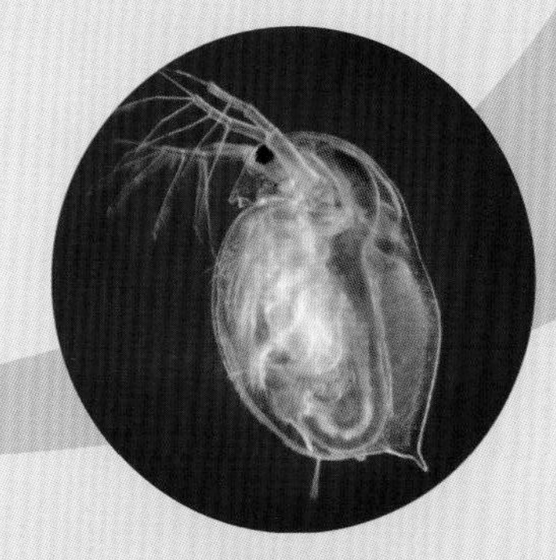
WATER FLEA

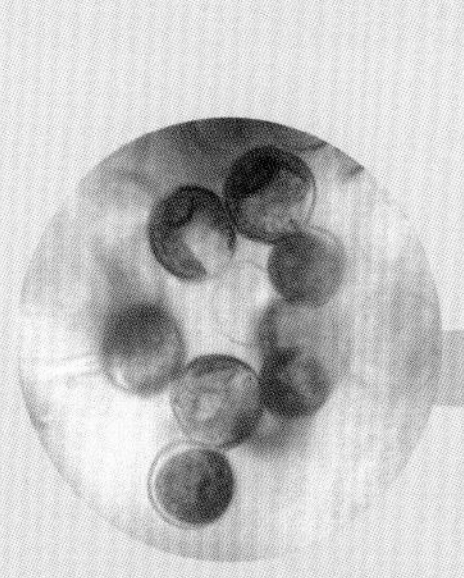
GREEN ALGAE

WIDOW SKIMMER (DRAGONFLY)

Dragonflies and their smaller damselfly cousins dart this way and that. Their iridescent colors gleam in the sun as they hunt for insect prey on the wing, grabbing them with their powerful front legs. These stunning creatures lay their eggs in the water or on plant stems under the water. Their larvae lie in wait for prey on the pond bottom. When they are ready, they leave the pond, emerge from their larval skins, unfold their wings, and begin to fly.

COMMON GREEN DARNER (DRAGONFLY)

SPINY BASKETTAIL (DRAGONLY)
EMERGING FROM LARVAL SKIN

BLUE PIRATE (DRAGONFLY)

EASTERN FORKTAIL (DAMSELFLY)

PAINTED TURTLE

Painted turtles warm themselves on a sunny log in the pond, while a ribbon snake waits for prey under some branches. Nearby, a tadpole hides among dead leaves at the bottom of the pond. Spots on its tail help it blend into its surroundings. This tadpole has already grown its legs; its tail will gradually get smaller and smaller until the tadpole transforms into a tiny gray tree frog. Spotted salamanders also lay their eggs in the beavers' pond. After the larvae grow up and sprout legs, they will move onto land.

SPOTTED SALAMANDER (LARVA)
SPOTTED SALAMANDER (ADULT)
GRAY TREE FROG (TADPOLE)
RIBBON SNAKE

GREAT BLUE HERON

An extraordinary variety of birds live in and around the beavers' pond. A great blue heron stands still as a statue, watching for fish—and is soon rewarded for its patience. Swallows soar and dip over the water, snatching insects from the air with their beaks. Ducks are drawn to the beavers' pond, where they can dabble for food on the quiet water instead of trying to feed in the running water of a creek. A couple of wood ducks swim by a family of black ducks, while a lone coot feeds on submerged plants.

TREE SWALLOW
AMERICAN COOT
AMERICAN BLACK DUCK
WOOD DUCK

CANADA GOOSE

Not only do animals live in and around the pond, some of them even use the structures built by the beaver. Sometimes muskrats settle into the beavers' family home—either while the beavers are living there or after they've moved out. The beavers don't seem to mind. Nearby, a pair of Canada geese choose the beavers' lodge as a protected island for their nest. A white-tailed deer uses the dam as a bridge to cross the water, and a duckling stands there, catching flies for dinner.

WHITE-TAILED DEER

MALLARD DUCKLING

MUSKRAT

Night is coming. This is when nocturnal animals become active. Frogs start to sing, creating a symphony, and a moose heads out of the forest to drink. Raccoons explore and scavenge. Hunters, such as the owl hooting high in a tree and the red fox emerging from its nearby den, seek their prey. It is also the time when beavers work their hardest.

MOOSE

RED FOX
BULLFROG
GREAT HORNED OWL
RACCOON

A beaver parent and kit grab a quick snack before slipping back into the water and swimming toward a stand of trees. The beavers' evening tasks await: There's a dam to be patched, trees to be cut, and food to be stored. The beavers work hard to keep their dam and lodge in good repair, unaware of how many animals and plants depend on their nightly work for survival.

If the beavers had never built this pond, many of the animals and plants that live here wouldn't have a home. After the beavers leave the pond, their handiwork will continue to play a crucial role in the ecosystem. Abandoned lodges serve as homes for small animals such as voles, other rodents, and snakes. Even without the beavers there to make repairs, the old dam will remain for a while, acting as a bridge over the creek for animals large and small. But eventually the dam will break, allowing the pond to drain and a meadow to form. The environment created by this keystone species will remain a gift to the diversity of life for years to come.

More About Beavers

Beavers are rodents, like mice, rats, and squirrels, but they are much bigger and live longer. While a large rat may weigh half a pound (.23 kg) and live for two years, a beaver can weigh more than seventy pounds (32 kg) and may live more than ten years. There are two species of beaver: the North American beaver and the Eurasian beaver. Eurasian beavers live in parts of Europe and Asia; North American beavers occur naturally as far south as northern Mexico and throughout most of the United States and Canada. The North American beaver has also been introduced to parts of Europe and South America, where they are now considered to be an invasive species.

Beavers have large front teeth (called incisors) that never stop growing. However, continual gnawing on wood helps to keep the teeth from getting too long. Beavers are most active at night and can hear well both in the air and underwater. A good sense of smell helps them find food and avoid predators. Their eyesight, although said to be poor, is actually quite good, especially underwater because of a clear membrane on the eye that acts like diving goggles. Beavers use their broad, scaly tails to steer when swimming and to slap the water to warn other beavers when a predator may be near.

Beavers often have large families called colonies. The mother beaver gives birth to several kits (baby beavers) in the springtime, while the beavers born the previous year or two are still at home, learning how to gather food and build a dam and lodge. When beavers reach two to three years old, they leave and may travel many miles to find a new home and a mate. If food or building materials become scarce, adult beavers may also leave the pond and construct a new dam and lodge elsewhere.

The lodge is a safe home for the beavers. It is virtually impossible for predators to get through the above-water mound of sticks and mud. The beavers make two underwater entrances to their lodge in case something tries to get in through one of them; this way they still have an escape route. The lodge has two levels inside, an upper one for sleeping and a lower one for eating and grooming near the underwater entrances. Beavers don't hibernate; instead, they dine all winter on the bark of logs and branches they stored in a pile near the lodge.

During the 1700s and 1800s, beavers in North America were almost trapped out of existence for their fur, which was made into fashionable hats. Luckily, they survived in small numbers here and there and have been expanding back into their old range. We now understand how important their activities are to the health of the natural environment. We've realized that the ponds made by beavers are essential habitats that provide a reliable source of water and food for many other living things. The beaver ponds and surrounding wetlands play an important role in replenishing groundwater, especially during droughts. The resulting healthy greenery can also help decrease the intensity and extent of wildfires and provide islands of life that can speed natural recovery of the landscape.

Even so, some states and Canadian provinces still allow beavers to be trapped and killed for their fur. People may also trap beavers because they don't like losing trees on their land or having the ponds flood their roads. Wildlife managers and the public who realize just how important these animals are to nature's diversity now have a number of ways to keep beaver ponds from flooding using fences and/or drainage tubes that can keep down the ponds' water levels. And in recent years, trained people have been helping beavers return to their old habitats by relocating them from places where they are not wanted to locations such as parks and nature preserves where they are safe to play their vital role as a keystone species.

Here are some resources for learning more about beavers and how to help protect them:

Beavers: Wetlands & Wildlife
www.beaversww.org

Beaver Institute
www.beaverinstitute.org

Worth a Dam
www.worthadam.org

See also Michael Runtz's book *Dam Builders: The Natural History of Beavers and Their Ponds*, which features his photos and lots of information about beavers, their ponds, and the plants and animals that benefit from them.

This book is dedicated to the organizations and individuals who work hard to bring this vital species back home where it belongs and can do its important work.
—D.H.P.

For Ann, for selflessly tolerating my frequent forays into the magical world of beavers.
—M.R.

Book design by Philip Krayna · Conifer Creative · www.conifercreative.com

For information, write to:
Web of Life Children's Books
P.O. Box 2726, Berkeley, California 94702

Published in the United States in 2019 by Web of Life Children's Books.

Printed in China through Four Colour Print Group, Louisville, Kentucky.
Production Date: 2018-11-29 Plant & Location: Printed in Guangdong, China Job / Batch #: 83084

Library of Congress Control Number: 2018963464
ISBN 978-1-970039-00-9

For more information about our books and the authors and artists who created them, visit:
www.weboflifebooks.com

Distributed by Publishers Group West/An Ingram Brand
(800)788-3123
www.pgw.com